ANIMALS AT RISK

Black-footed Ferrets

by Rachel Grack

BLASTOFF! READERS 2

BELLWETHER MEDIA • MINNEAPOLIS, MN

Blastoff! Readers are carefully developed by literacy experts to build reading stamina and move students toward fluency by combining standards-based content with developmentally appropriate text.

Level 1 provides the most support through repetition of high-frequency words, light text, predictable sentence patterns, and strong visual support.

Level 2 offers early readers a bit more challenge through varied sentences, increased text load, and text-supportive special features.

Level 3 advances early-fluent readers toward fluency through increased text load, less reliance on photos, advancing concepts, longer sentences, and more complex special features.

★ Blastoff! Universe

Reading Level

Blastoff! Beginners — Grade K

Blastoff! Readers — Grades 1–3

Blastoff! Discovery — Grade 4

This edition first published in 2023 by Bellwether Media, Inc.

Library of Congress Cataloging-in-Publication Data

Names: Koestler-Grack, Rachel A., 1973- author.
Title: Black-footed ferrets / Rachel Grack.
Description: Minneapolis, MN : Bellwether Media, 2023. | Series: Blastoff! readers: animals at risk | Includes bibliographical references and index. | Audience: Ages 5-8 | Audience: Grades 2-3 | Summary: "Relevant images match informative text in this introduction to black-footed ferrets. Intended for students in kindergarten through third grade"-- Provided by publisher.
Identifiers: LCCN 2022000425 (print) | LCCN 2022000426 (ebook) | ISBN 9781644877098 (library binding) | ISBN 9781648347559 (ebook)
Subjects: LCSH: Black-footed ferret--Juvenile literature. | Black-footed ferret--Conservation--Juvenile literature.
Classification: LCC QL737.C25 K64 2023 (print) | LCC QL737.C25 (ebook) | DDC 599.76/629--dc23/eng/20220113
LC record available at https://lccn.loc.gov/2022000425
LC ebook record available at https://lccn.loc.gov/2022000426

Editor: Kieran Downs Designer: Brittany McIntosh

Printed in the United States of America, North Mankato, MN.

Table of Contents

Masked Mammals

Black-footed ferrets are long, thin **mammals**. They have black fur around their eyes. It looks like a mask!

They live in
grassland **habitats**.

Black-footed ferrets once lived across the **Great Plains**. Today, they are greatly **endangered**.

People caused most of their troubles.

Black-footed Ferret Range

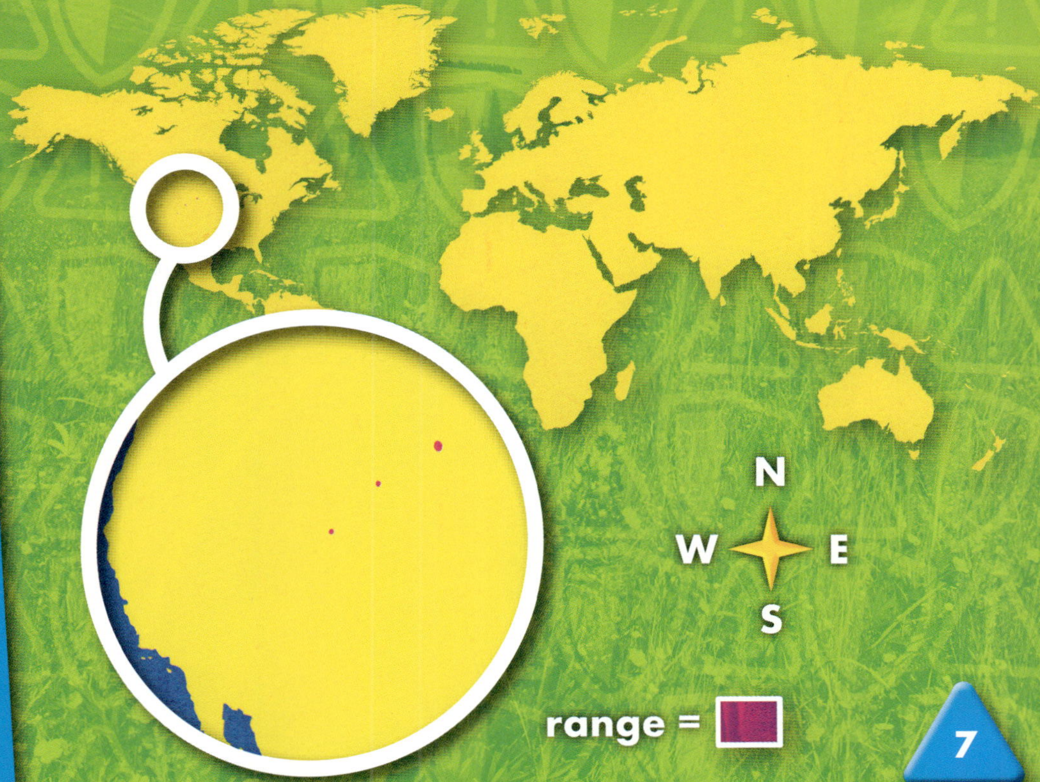

In Danger!

Black-footed ferrets eat prairie dogs. They live in prairie dog tunnels.

But farmers got rid of prairie dogs on their land. Sickness also killed prairie dogs. Ferrets lost homes and food.

prairie dogs

Black-footed Ferret Stats

Least Concern	Near Threatened	Vulnerable	Endangered	Critically Endangered	Extinct in the Wild	Extinct

conservation status: endangered

life span: up to 3 years

People cleared **prairies** to build towns and farms. Black-footed ferrets lost their habitat.

People thought the ferrets were **extinct**. But a small group was found.

Threats

1 people need farmland

2 people plant fields

3 black-footed ferrets lose homes

Save the Black-footed Ferrets!

Black-footed ferrets keep grasslands healthy. They control the number of prairie dogs.

Ferrets are also food for other animals.

The World with Black-footed Ferrets

1

more black-footed ferrets

2

healthy number of prairie dogs

3

healthy grasslands

Wildlife groups raise black-footed ferrets. Young ferrets stay safe and healthy.

The animals' numbers are growing. More can return to the wild.

Some states set aside
land for prairie dogs.
Later, wildlife workers move
black-footed ferrets there.

These places become healthy grassland habitats.

Wildlife workers **vaccinate** black-footed ferrets.

They also drop pills in prairie dog tunnels. This stops sicknesses from spreading.

Groups work with farmers to save prairie dog homes. **Donations** help wildlife groups care for black-footed ferrets.

Together, everyone can save these prairie animals!

Glossary

donations—gifts for a certain cause; most donations are money.

endangered—in danger of dying out

extinct—completely gone from the wild

Great Plains—a region of flat or gently rolling land in the central United States

habitats—places and natural surroundings in which plants or animals live

mammals—warm-blooded animals that have backbones and feed their young milk

prairies—large grasslands that are mostly flat and treeless

vaccinate—to give a powerful medicine that helps increase protection against a harmful disease

To Learn More

AT THE LIBRARY

Duling, Kaitlyn. *Prairie Dogs*. Minneapolis, Minn.: Bellwether Media, 2021.

Rossiter, Brienna. *Saving Earth's Animals*. Lake Elmo, Minn.: Focus Readers, 2022.

Stuckey, Rachel. *Bringing Back the Black-Footed Ferret*. St. Catharines, Ont.: Crabtree Publishing Company, 2020.

ON THE WEB

FACTSURFER

Factsurfer.com gives you a safe, fun way to find more information.

1. Go to www.factsurfer.com.

2. Enter "black-footed ferrets" into the search box and click 🔍.

3. Select your book cover to see a list of related content.

Index

The images in this book are reproduced through the courtesy of: Kerry Hargrove/ Alamy Stock Photo, front cover, p. 22; turtix, front cover (back top), pp. 3, 23; Rui Serra Maia, front cover (back bottom); USDA Photo/ Alamy Stock Photo, p. 3; Laura Romin & Larry Dalton/ Alamy Stock Photo, pp. 4, 18; Kerry Hargrove, pp. 5, 8-9, 12, 13 (top left), 17; Kerry Hargrove | Dreamstime.com, pp. 6, 20-21; AB Photographie, p. 8; Gorlov-KV, p. 10; Emily Marie Wilson, p. 11 (top left); oticki, p. 11 (top right); Steam visuals, p. 11 (bottom); Natallia Kuzmina, p. 13 (top right); Jason Patrick Ross, p. 13 (bottom); The Washington Post/ Contributor/ Getty Images, p. 14; Nature and Science/ Alamy Stock Photo, pp. 15, 19, 20; Images by Dr. Alan Lipkin, p. 16.